Mohammed KOUIDRI
Kaouthar Chettfour

Problems of the cultivated olive tree south of the Saharan Atlas (Laghouat)

Mohammed KOUIDRI
Kaouthar Chettfour

Problems of the cultivated olive tree south of the Saharan Atlas (Laghouat)

Diseases and pests in selected orchards

ScienciaScripts

Imprint
Any brand names and product names mentioned in this book are subject to trademark, brand or patent protection and are trademarks or registered trademarks of their respective holders. The use of brand names, product names, common names, trade names, product descriptions etc. even without a particular marking in this work is in no way to be construed to mean that such names may be regarded as unrestricted in respect of trademark and brand protection legislation and could thus be used by anyone.

Cover image: www.ingimage.com

This book is a translation from the original published under ISBN 978-620-2-26955-1.

Publisher:
Sciencia Scripts
is a trademark of
Dodo Books Indian Ocean Ltd. and OmniScriptum S.R.L publishing group

120 High Road, East Finchley, London, N2 9ED, United Kingdom
Str. Armeneasca 28/1, office 1, Chisinau MD-2012, Republic of Moldova, Europe
Managing Directors: Ieva Konstantinova, Victoria Ursu
info@omniscriptum.com

Printed at: see last page
ISBN: 978-620-8-39399-1

CONTENTS

INTRODUCTION 2

CHAPTER 1 4

CHAPTER 2 12

CHAPTER 3 18

CHAPTER 4 34

CHAPTER 5 44

CONCLUSION 49

REFERENCES 50

INTRODUCTION

Olive growing (Olea europaea, L.) is one of the oldest and most widespread activities in the arid and semi-arid areas of the Mediterranean basin, mainly due to its great adaptability to water deficit conditions and its nutritional value (Connor and Fereres, 2004; Fernández, 2014). The Algerian government introduced a national olive cultivation plan (PNO) in 2000, the aim of which was to increase the extension of olive groves (Argenson, 2008). However, the success of this initiative has been hampered by monitoring and maintenance problems. At present, the olive tree suffers from a number of problems that affect both its production and its numbers (El Hadrami and Nezha, 2001), the most important of which are edaphoclimatic conditions (salinity, drought and silting) (Loussert and Brousse, 1978), bacterial diseases (Assawah and Ayat, 1985), fungal diseases (Bellahcene, 2004; Bellahcenne et al, 2005a, Bellahcenne et al., 2005b), peacock eye (Guechi and Girre, 2002) and, above all, a number of pests: black scale (Loussert and Brousse, 1978), olive fly (Gaouar, 1996), olive moth (Gaouar, 1996) and European starling (Sturnus sturnus) (Bellatrèche, 1986; Rahmouni-Berrai, 2009). In the same context, other important works include those by Al Ahmed and Al Hamidi (1984), Alford (1994), Guario and La Notte (1997), Alvarado (1999), Coutin (2003) and Duriez (2001).Our main objective is to analyse the current state of health of one of the pioneering projects launched in the southern region of Laghouat, namely the Bennacer Ben Chohra olive plantation. This project, launched in 2007, is suffering from a number of problems that are hampering the proper development of its trees. We tried to identify the problems and the main pests that can be observed during the spring period. The first is a review of the literature on the olive tree, its enemies and the methodology adopted during this study. The second part concerns the results obtained, which will then be discussed in the light of the available literature on the species under study. The work ends with a conclusion and outlook.

PART 1

BIBLIOGRAPHICAL SUMMARY

CHAPTER 1

GENERAL INFORMATION ON THE OLIVE TREE

This chapter deals with the bibliographical study of the olive tree, followed by a study of some of the pests that cause damage to this tree.

1. Bibliography on the olive tree

The bibliographical study of the olive tree covers its history, systematics, characteristics, requirements, varieties and distribution throughout the world and in Algeria.

1.1. History

According to Henry (2003), historians and archaeologists are not unanimous about the country of origin of the olive tree, but this tree undoubtedly found natural conditions in the Mediterranean, climatic constraints, to which it adapted perfectly, so the expansion of the olive tree is linked to the establishment of the Mediterranean climate. The Mediterranean climate has evolved gradually since 10,000 BC, with the olive tree first establishing itself in the eastern Mediterranean, then spreading over several millennia to the west and north of the Mediterranean basin (Amouretti and Comet, 2000). Evidence of this can be found as far back as the fourth millennium BC, and according to some even as far back as 10,000 years ago (Artaud, 2008). This species originates from Asia Minor or Crete. The first traces of this tree date back to 37,000 BC, on fossilised leaves discovered on the islands of Santorini in Greece (Henry, 2003). Biological studies show that the wild olive tree existed in the Sahara around 11,000 years before our era. The latest analyses of the pollens of various deciduous and dominant trees seem to show that this climatic change developed around 8,000 years BC, in south-eastern Spain, moving slowly northwards. As early as 3,000 BC, the olive tree was cultivated in Egypt, Syria, Palestine and Phoenicia (Henry, 2003). Around 1600 BC, the Phoenicians spread the olive tree throughout Greece. From the 6th century BC, its cultivation spread throughout the Mediterranean basin in via Libya, Tunisia, Sicily and then Italy. During their conquests, the Romans continued to spread the olive tree to all the countries along the Mediterranean coast (Henry, 2003).

1.2. Systematics

According to Iguergaziz (2012), the systematics of the olive tree is as follows:
Kingdom: Plantae
Phylum: Spermaphytes

Subphylum: Angiosperms Class: Magnoliopsida (Dicotyledons) Subclass: Asteridae
Order: Srophulariales Family: Oleaceae Genus: Olea
Species: Olea europaea (Linné, 1753)

1.3. Characteristics of the olive tree

The morphological characteristics of the olive tree according to Sekour (2012) are normally distinguished by a short trunk, dark and deeply fissured bark, rough and tortuous, and a broad, branched head that can reach up to 4 to 5 metres (Photo 1).

Photo 1: Olive tree in the study orchard (original)

According to Loussert and Brousse (1978) the leaves of the olive tree are entire and lanceolate, arranged on the branches with a short petiole. Like all Oleaceae, they are opposite. The leaves are evergreen, lasting an average of 2 to 3 years. Their size varies between 3 and 8 cm long and 1 to 2.5 cm wide, depending on the variety. According to Henry (2003) the flowers of the olive tree are small and white in colour formed by a tetrameric flower (four petals), an oval calyx, two very short filament stamens, and a rounded ovary which bears a style this ovary contains two ovules. According to Djadoun (2011) the fruit of the olive tree is very rich in lipids and is ovoid in shape, 2 to 4 cm long. Its root system is very dense, so it is firmly anchored in the soil and can withstand wind, drought and erosion. Sometimes it has large bulges, which are reserves that enable it to cope with climatic variations (Artaud, 2008). To perform these functions as well as possible, the root system needs a large volume of soil to explore, containing oxygen, water and assimilable nutrients (COI, 2007). According to Kasraoui (2010), the final appearance of the root system depends on the physicochemical characteristics of the soil and the depth of texture and structure.

1.3.1. Olive tree development cycle

According to Loussert and Brousse (1978), the olive tree goes through four phases, the first of which is the juvenile period, which extends from sowing to the first flowering over a period of 4 to 9 years. A juvenile plant can be distinguished by its morphology. It has a very bushy growth habit, numerous branches with more or less short anticipated ramifications and small, broad leaves. The second is the production period, which lasts from 12 to 50 years, during which it begins to produce while continuing to grow. The third is the adult period, which lasts from 50 to 150 years, when the tree is fully mature and producing abundantly. Finally, there is the senescence period, after 150 years, when the trunk begins to hollow, the tree loses some of its bark and production declines.

1.3.2. Annual vegetative cycle

According to Loussert and Brousse (1978), winter rest extends from November to February. At this stage, the terminal bud and axillary eyes are in vegetative rest. Spring awakens between March and April, when new terminal shoots appear and the axillary buds open. Flowering takes place between May and June, with the formation of flower clusters, followed by the setting of young fruit, and then the enlargement of the fruit, which reaches 8 to 10 cm in length. In October, the fruit ripens and is enriched with oil.

1.3.3. Requirements of the olive tree

According to Labaali (2009), the olive tree fears humidity, but can withstand exceptional droughts (application of thirty to forty litres of water, once or twice in July and August, and only in the first year after planting and 450 and 600 mm/year, production is possible provided the soil has water retention capacity, etc.). or the density of the plantation is lower). High humidity, hail and spring frosts are all unfavourable factors for flowering and fruiting.

According to Boutkhil (2012) the areas where olive trees are most widespread are characterised by mild winters, temperatures rarely below 0o C and dry summers with high temperatures.

According to Duriez (2004), edaphic requirements show that the olive tree's root system extends preferably in the first 50 to 70 cm of soil, with the roots being able to reach a depth of up to one metre to seek additional water. This is why the soil must be adapted in terms of texture, structure and composition to a depth of at least one metre. Loussert and Brousse (1978) point out that the most suitable soils for olive trees are those characterised by a balance of sand, silt and clay. Predominantly sandy soils have a low water and mineral retention capacity, but allow good aeration of the land and are an advantage for olive trees when water is available, provided that appropriate fertilisation is carried out to meet nutritional requirements in terms of mineral elements. Quantities of clay should not be excessive, as they could act as an obstacle to air circulation and soil movement. According to Gazeau (2012), the olive tree prefers relatively poor soil to very fertile soil. It should not be planted in very fertile, deep soils. The aim of optimum fertilisation of the olive tree is to produce a regular harvest, good vegetative development and good resistance to the cold in winter.

Density and spacing are other important choices that are conditioned by variety, soil and climate. For the same density, square or almost square spacing gave better results than rectangular spacing (CIHEAM, 1988). The same author points out that to determine the planting density, the final development of the tree and its rate of growth must be taken into account. The planting distance must allow the foliage to capture the maximum amount o f solar energy, without mutual shading between neighbouring trees. At In olive growing, the distance between trees in the row varies between 5 and 7 m depending on the variety (COI, 2007).

1.3.4. Varieties of olive trees grown around the world

According to Loussert and Brousse (1978) the dominant varieties worldwide are those found in Tunisia as oil olives (Chemlali and Chetoui) and table olives (Marsaline). Other varieties are found in Spain as oil olives (Hajiblanca and Verdal) and table olives (Manzanille and Gordal-sevillana). In Italy we find the oil olive (Moraiolo and Leccino) and the table olive (Ascolona Tenera and Santa Caterina).

1.3.5. Olive varieties grown in Algeria

According to Iguergaziz (2012) the Kabylie varieties of oil olive are Chemlal, Limli and Bouchouk. According to Loussert and Brousse (1978) the table olive is Sigoise, Adjeraz or Azeradj. Other varieties have been introduced, such as the Spanish Corncabra and the French Verbal.

1.3.6. World olive-growing area and production

The current olive-growing heritage is estimated at approximately 1,000 million trees occupying a surface area of 10 million ha, 98% of which is located in the Mediterranean basin, 1.2% on the American continent, 0.4% in East Asia and

0.4% in the countries of the Pacific Ocean. Algeria, a Mediterranean country with a climate conducive to olive growing, ranks after Spain, Italy, Greece and Tunisia respectively as the world's largest olive oil producers (IOC, 2015) (Fig.1).

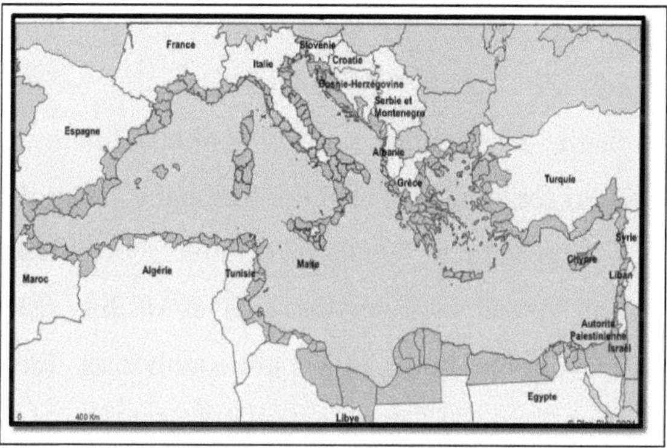

Figure 1: Natural distribution of the olive tree worldwide

1.3.7. Olive tree distribution in Algeria

In Algeria, there are around 281,000 ha of olive groves, to which must be added 110,000 ha that have gradually come into production since 2007 (MADRP, 2013) (Fig. 2).

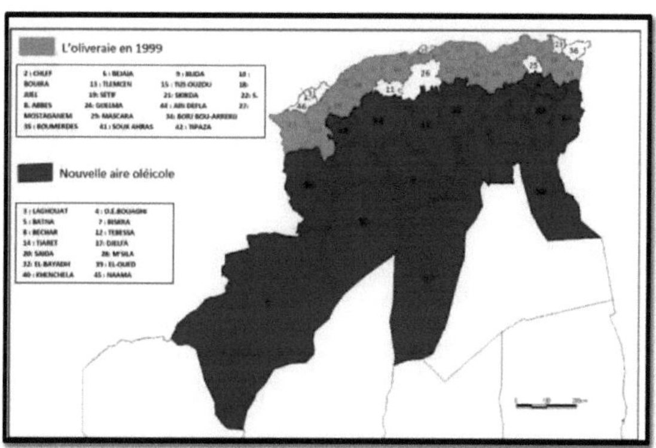

Figure 2 : Olive-growing map of Algeria (ITAFV, 2008).

The Algerian olive grove is spread over three major olive-growing areas, the western region representing 31,400 ha spread over 5 wilayas (Tlemcen, Ain Temouchent, Mascara, Sidi Bel Abbes and Relizane). This western zone represents 16.4 of the national olive orchard. The central region covers an area of 110,200 ha, divided between the wilayas of Ain Defla, Blida, Boumerdès, Tizi Ouzou, Bouira and Bejaia. This central area represents 57.5% of the national olive orchard. The central Kabylia region (Bouira, Bejaia and Tizi Ouzou) alone accounts for almost 44% of the national olive-growing area.

The Eastern region is made up of 49,900 ha of olive groves, representing 26.1% of the national total, spread between the wilayas of Jijel, Skikda, Mila and Guelma (Sekour, 2012). The range of varieties authorised for production and multiplication, according to the Centre national de contrôle et certification des semences et plants, covers a total of 46 olive varieties; however, the varieties most multiplied during the 2015/2016 season were limited to just 14 varieties. The varieties in question are according to (CNCC, 2015): Sigoise; Chemlal; Azeraj; Tefah; Manzanille; Aberquina

Sevillane; Ferkani; Belgentieroise; Bouchouk Soummam; Blanquette de

Guelma; Rougette; Hamra; Grosse du Hamma.

1.3.8. Soil and climate requirements of olive trees

1.3.8.1. Climatic requirements

a) Temperature

The olive tree grows in countries with a Mediterranean climate, where temperatures vary between 16 and 22°C (average annual temperature). It likes light and heat, tolerates high temperatures very well, even in a dry atmosphere, and is not afraid of sunstroke (Tahraoui, 2017). According to Boutkhil (2012), the areas where olive trees are most widespread are characterised by mild winters, temperatures rarely below 0°C and dry summers with high temperatures. It also fears the cold, and sub-zero temperatures can be dangerous, particularly if they occur at flowering time (Hannachi et al., 2007).

b) Rainfall

Winter rainfall allows the soil to store water reserves. Autumn rainfall in September and October encourages fruit to grow and ripen. Rainfall should not be less than 220 mm per year (Benrachou, 2013), a low figure which shows that the olive tree tolerates drought well. In fact, it is content with low rainfall, the lowest of all fruit species. The period from 15 July to 30 September is very important for fruit development. If this period is too dry, the fruit falls prematurely and the yield is considerably reduced. This is why irrigation is sometimes necessary to avoid this accident (Benrachou, 2013).

1.3.8.2. Soil requirements

The olive tree has no particular requirements in terms of soil quality. It is reputed to be content with poor soil, whether clayey or, on the contrary, light or stony, but it must be deep enough to allow the roots to nourish the tree by exploring a sufficient volume of soil. Olive trees do not like soil that is too wet. The soil must have a high nitrogen content (Hannachi et al., 2007).

CHAPTER 2

THE ENEMIES OF THE OLIVE TREE

2.1. Main olive tree diseases

There are 2 types of disease: abiotic and biotic:

2.1.1. Diseases of abiotic origin

There are several diseases of abiotic origin in olive trees (Tab. 1):

Table 1: Different types of problems affecting olive trees (Saad, 2009)

Type incidents	Factors favourable	Manifestation of symptoms
Climatic accidents	Gel Sunstroke burns	Fall of leaves ; necrosis of young bark, parasitic infection. Damage to young plantations, to the tissues of the on trunks and carpenters
Weather accidents	Heavy snow Hail High winds	Foliage breakage on fruit harvesting, breakage and injuries to young bark, spread of tuberculosis. Breakage of carpentry, reduced harvest
Root asphyxia	Soils with too much moisture and clay	Yellowing (chlorosis), defoliation stoppage of vegetative growth, early fruit drop
Dietary chlorosis	Element deficiencies Indispensables (nitrogen, limestone and Cl - and Na+ ions	Serious physiological disorders in plants

2.1.2. Diseases of biotic origin

The olive tree, like other fruit trees, is often attacked by a multitude of bio-aggressors (Bellahcene, 2004), including 110 insect species, 100 nematode species, 90 fungal species, 13 arachnid species, 13 viruses, 5 bacterial species, 4 mosses, 3 lichens and 3 angiosperms (Faustino de Andres, 1965; Sasanelli, 2009). 11 nematodes, 110 insects, 13 arachnids, 5 birds and 4 mammals (Maillard, 1975 in Gaouar, 1996). It is also home to a wide variety of fauna, including plant-eating species that can cause significant damage, both in terms of quantity and quality (Rahmani, 1999).

2.1.2.1. Viral diseases of the olive tree

With regard to viral diseases, most viruses, with the exception of cryptoviruses, are associated with more or less serious damage to the plants they parasitise, resulting in quantitative and/or qualitative crop losses (Clara et al., 1997). The Manzanillo variety grown in Palestine has been affected by a Spherosis virus (Lavee and Tanne, 1984). In Italy, Savino and Gallitelli (1983) showed that a virus that attacks cherries also causes leaf curl in olive trees. Other authors have reported viral symptoms in olive crops in Greece (Barba, 1993; Kyriakopoulos, 1993). There are also other diseases that are highly responsive and affect the olive tree, such as :

2.1.2.2. Olive scab (Peacock eye) (Cycloconium Oleaginu Cast.)

This is the best-known disease of the olive tree. In addition to Mediterranean countries, it is found in South Africa, Eritrea, the United States and Chile. It is almost exclusively a parasite of Olea europaea, although some authors have reported attacks by a strain of the parasite on the genera Phyllirea and Ligustrum (Guechi and Girre, 2002) (Photo 2-G).

2.1.2.3. Olive wilt (Verticillium dahliae Kleb.)

This pathogen is very widespread, attacking a large number of species, both woody and herbaceous. On olive trees, it was first described in Italy in 1946 and has been observed in various countries around the Mediterranean basin, notably Spain, France, Greece and Turkey, but also in Asia Minor, Syria and the United States (California) (Bellahcene, 2004; Bellahcene et al., 2005a, 2005b) (Photo 2-F).

2.1.2.4. Bacterial canker of the olive tree :

It is an infectious disease caused by a bacterium, P. savastanoi, which was first reported in the 4^{éme} century by the Greek, Theophrastus. The pathogen appears to have been disseminated with olive plants (Olea europaea subsp. europaea) and then spread to many regions of the world. This bacterium was isolated by Luigi Savastanoi (Bradbury, 1986).

The current name is Pseudomonas syringae pv. savastanoi (Gardan et al., 1992). This bacterium is considered to be the only pathogen responsible for the formation of bacterial knots (necrosis) in olives (Philippe, 2007).

2.1.2.5. Fumaginia (Capnodium ssp.; Alternaria ssp.):

Fumaginia or "olive black" is a disease carried by various fungi that develop on the sugary substances in the honeydew secreted by sap-sucking insects (black olive scale, psyllid). The leaves are covered with a kind of black dust resembling soot, preventing the tree from breathing and condemning it to death by asphyxiation (Photo 2-H).

2.1.2.6. Cercosporiosis of olive trees:

In 1944, the disease was described for the first time in America. In 1968, an epidemic of Cercosporiosis broke out on olive trees, with the pathogen attributed to Mycocentrospora cladosporioides in Greece (Freeman et al. al., 1998; Agosteo et al., 2000). Cercosporiosis can cause serious damage, particularly in hot-summer regions of Europe, and the disease reduces yields. Other species are pests of the Olive tree such as:

2.1.2.7. Olive fly (Dacus oleae) :

According to I.N.P.V. (2009) the olive fly Dacus oleae is the most worrying pest for olive growers, causing damage to fruit that can amount to up to 30% of fruit being damaged and unusable. Attacks by the fly also lead to a deterioration in the quality of the oil, causing an increase in acidity (Photo 2-B).

2.1.2.8. Olive moth (Prays oleae) :

According to Jardak et al (2000), the moth is the first major pest to be observed under the leaves of olive trees in March. This pest can cause significant crop losses. Recognition of this pest is essential to ensure appropriate and effective control (Photo 2-D).

2.1.2.9. Olive black scale (Saissetia oleae) :

According to Loussert and Brousse (1978), Saissetia oleae is an insect of the Sternorhynches family. Like the aphid or psyllid, it is not specific to the Olive tree, as it also lives on other plants, in particular the Pink Laurel. As an adult, it measures around 5 mm long and 4 mm wide. It looks like a black half-sphere stuck to the inside of leaves, but especially on young stems that are a year or two old (Photo 2-C).

2.1.2.10. Olive psyllid (Euphyllura olivina):

This pest is small (2 mm to 2.5 mm) and dark grey in colour. The adults overwinter and the spring eggs are laid in March-April on the face of the plant. The larvae produce abundant honeydew (Coutin, 2003) (Photo 2- A). Other beetles are present in orchards, such as Otiorhyncus cibricollis (Photo 2-I), which is known for its damage to leaves.

2.1.2.11. European starling (Sturnus vulgaris) :

The European Starling (Sturnus vulgaris) belongs to the class of birds, the order Passeriformes, the family Sturnidae and the genus Sturnus (Berlioz, 1950). The French name Étourneau sansonnet is not universal (Cerny et Drachal, 1993), and it is commonly referred to by the English as the "European starling, Common starling and English starling" (Masterson, 2007). It is known for its migratory ethology, which distinguishes it from another species, the common starling (Sturnus unicolor) (Etchecopar and Hue, 1964). Although these two species share a great morphological similarity (Pascal and Peris, 1992) (Photo 2-F).

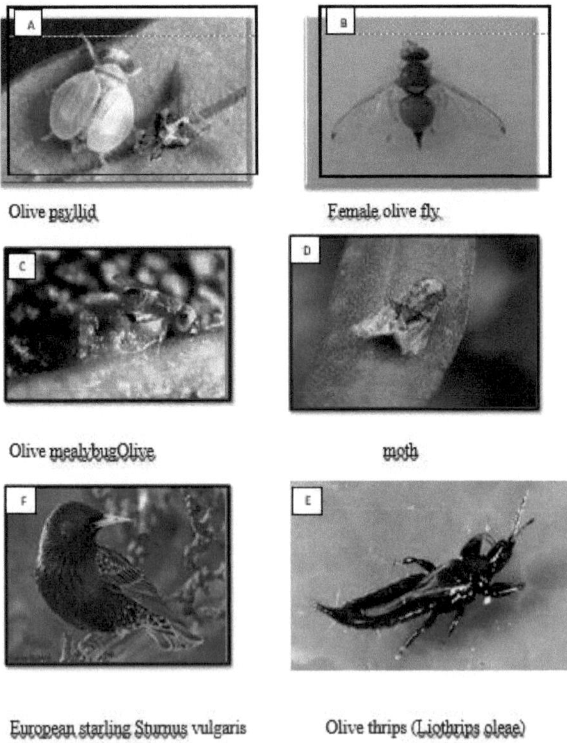

Photo 2: Various olive pests and diseases (INPV, 1994).

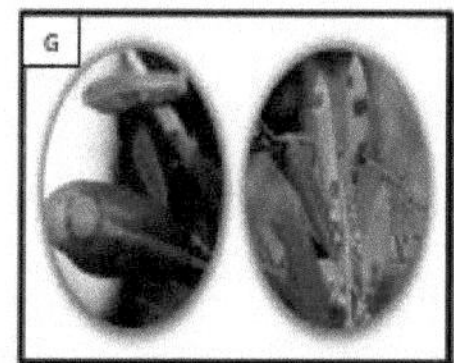

Damage caused by Verticillium dahliae (CTO, 2011)

Peacock eye on leaves (CTO, 2011)

Fumaginia on leaves (CTO, 2011)

Olive longhorned beetle (original)

(Continued) **Photo 2:** Various olive pests and diseases (INPV, 1994).

CHAPTER 3
MATERIALS AND METHODS

As part of the development of marginal land in arid and semi-arid zones, projects have been launched by the State to use several species of fruit trees. These trials began with the introduction of the olive tree, the atlas pistachio and other species. At present, very few studies have been carried out on the olive tree in our region, whether wild or cultivated. The present work is part of this effort to evaluate the pioneering sites of this development. The main objective of this study is to assess the state of health of the olive orchard studied and to identify the various problems affecting this species in our region.

3.1. Presentation of the study region :

Situated in the heart of the country, 400 km south of the capital Algiers, the region of Laghouat, by virtue of its geographical position and climatic characteristics, is in the central part of the country's nine pastoral wilayats. It covers an area of 25,052 km^2 . It is criss-crossed by the Saharan Atlas chain and bordered to the north by the wilaya of Tiaret, to the south by the wilaya of Ghardaïa, to the east by the wilaya of Djelfa and to the west by the wilaya of El Bayadh.

3.2. Natural features of the study region :

The physical and natural characterisation of the study area was based on the differentiation of the natural facies across the extent of the area along a north-south and east-west gradient. The identification of natural areas and the determination of their characteristics were inspired by the results of the division of the national territory into homogeneous agro-ecological zones (BNEDER, 2015) as follows:

3.2.1 . High Plateaux zone :

The high plains of the interior, which make up the high plateau zone, stretch between the tell and the pre-Saharan mountains of the Saharan atlas. They are on average 200 km wide and between 1000m and 1200m high.

3.2.2 . Saharan Atlas zone :

The Saharan Atlas closes off the high steppe plains to the south, consisting of a series of folded limestone and marl ranges, stretching from the Moroccan border at Naâma and El-Bayadh to Biskra in the east.

3.2.3 . Southern foothills of the Atlas :

This area is the southern extension of the Saharan atlas and is made up of foothills of high plains with a Saharan character.

3.3 . Climatic characteristics :

The climate in the study region is pre-Saharan Mediterranean, with a hot summer dry season alternating with a winter season with little rainfall. The decrease and increased irregularity of rainfall, the rise in temperatures and the length of the summer dry period make it more difficult for plants to develop with a water deficit (Le Houérou, 1996).

The climatic data used in our study are those provided by the Office national de météorologie (O.N.M) in Laghouat. These data were collected over a 15-year period, from 2004 to 2018.

3.3.1 . Precipitation :

Rainfall is of prime importance, as the amount of water reaching the ground is normally what supplies the trees. Rainfall varies according to three parameters (Kadik, 2005):

- Latitude, since rainfall decreases from north to south..;
- The longitude at which rainfall decreases from east to west;
- Rainfall increases with altitude.

Analysis of Figure 3 shows that inter-annual variability is significant, ranging from 66.8 mm for the driest year, 2017, to 285.2 mm for the wettest year, 2010, giving a difference of 218.4 mm. The average annual rainfall for the period under consideration is 167.83 mm/year.

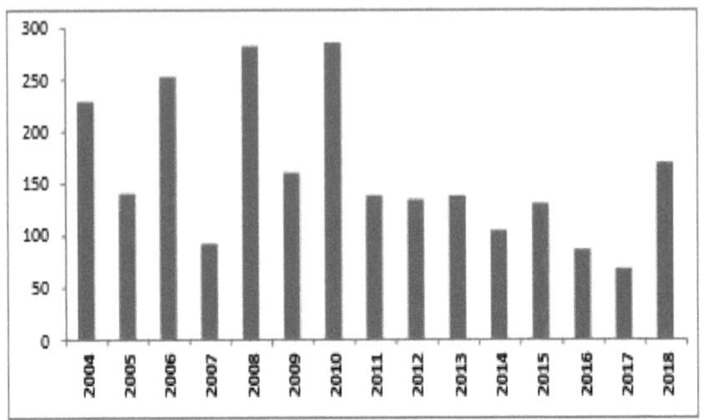

Figure 3: Interannual variability in mm of rainfall in the Laghouat region (2004-2018)

Average monthly rainfall shows that September is the wettest month with a value of 26.78 mm and July is the driest with a value of 6.77 mm (Fig. 4).

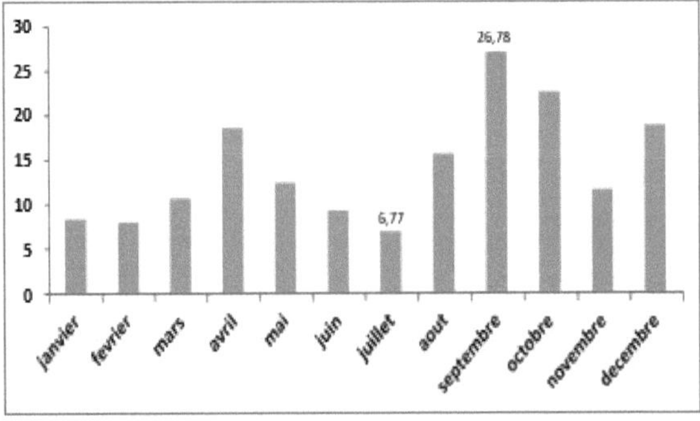

Figure 4: Variation in average monthly rainfall in the Laghouat region (2004-2018).

3.3.2 Temperature :

Among the factors limiting the presence and distribution of the olive tree, temperature is one of the most decisive. Each species has a minimum or maximum threshold that allows it to survive, beyond which the survival of the species may be compromised (Bentouati, 2006). In our area, the olive tree is exposed to temperatures that exceed its tolerance thresholds, particularly in summer. This temperature induces water stress for the species. The average annual temperature is 18.96°C, the hottest month of the year is July at 30.72°C and the coldest month is January at 8.77°C. The average maximum temperature of the hottest month "**M**" is 37.22°C and the average minimum temperature of the coldest month "**m**" is 2.3°C (Tab. 2).

Table 2: Average monthly temperatures in the Laghouat region (2004-2018)

Month	Jan	Feb	March	Apr	May	June	July	August	Seven	Oct	Nov	Dec
T°average	8,77	9,6	13,49	17,72	22,36	27,34	30,72	30,54	25,56	20,17	12,49	8,83
T° max	15,25	15,63	20,08	24,6	29,32	34,84	37,22	37,88	31,88	26,42	18,26	14,56
T°min	2,3	3,58	6,9	10,85	15,4	19,84	24,22	23,2	19,24	13,92	6,72	3,11

Source : O.N.M ,2019

The annual temperature range (2004-2018) is 21.95°C. This represents the difference between the warmest and coldest months (Tab. 2).

3.3.3 Wind :

Wind is a very important ecological factor that plays a role in seed dispersal. However, winds cause variations in temperature and humidity and have a harmful effect on plant behaviour. The frequency of wind acts on plants, especially on their aerial parts, by increasing evapotranspiration (Ozenda, 1983). The prevailing winds in the region are north-westerly. Generally in winter, they bring autumn and winter rain. Siroccos are most common throughout the year. They are dry and hot. In winter, they are fairly rare and are due to depressions affecting the Algerian coast. In summer, they are due to the Saharan influence. Siroccos play an essential role in the evaporation and drying out of olive groves in our region, and exacerbate water stress.

The maximum wind speed was recorded in April, at 4.08 m/s (Fig. 5).

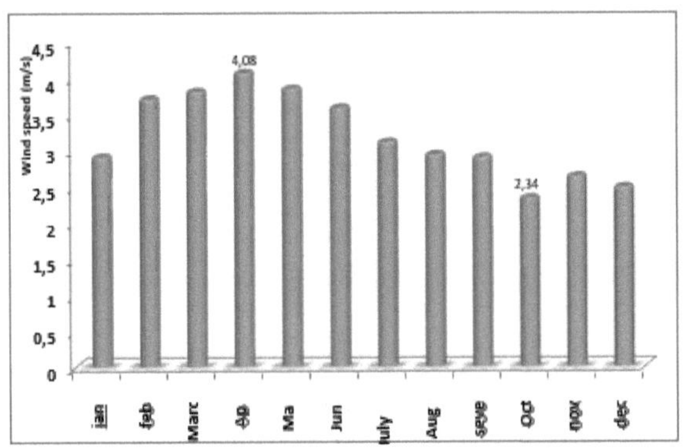

Figure 5: Evolution of the monthly average wind speed for the Laghouat region (2004-2018).

The strongest winds were recorded in 2010, with an average speed of 4.11 m/s, and the weakest in 2005, with an average speed of 2.09 m/s (Fig. 6).

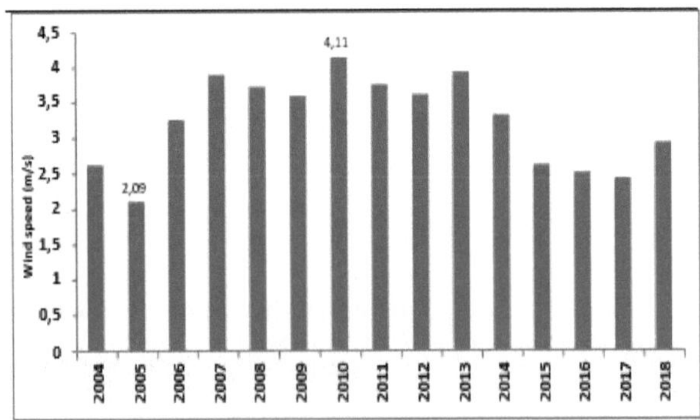

Figure 6: Interannual variation in wind speed in the Laghouat region (2004-2018)

3.3.4 Humidity :

The average annual relative humidity is 45.35%, with a minimum in July of 25.13%. Its maximum is recorded in December, with a value of 63.4% (Fig. 7).

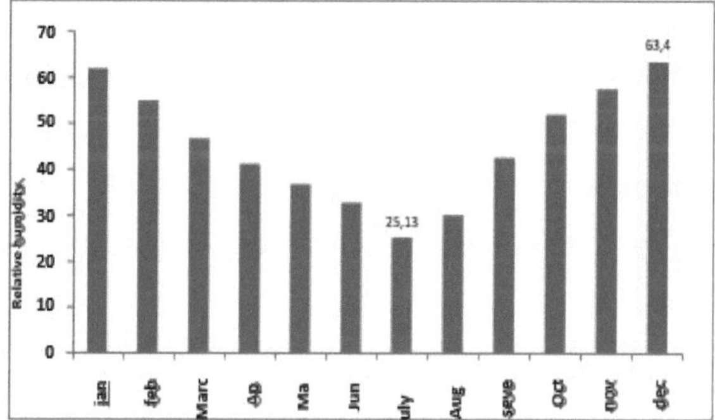

Figure 7: Change in average monthly relative humidity for the Laghouat region (2004-2018)

The wettest year was 2006 with 53.58% and the driest year was 2016 with an average of 28.91% (Fig. 8).

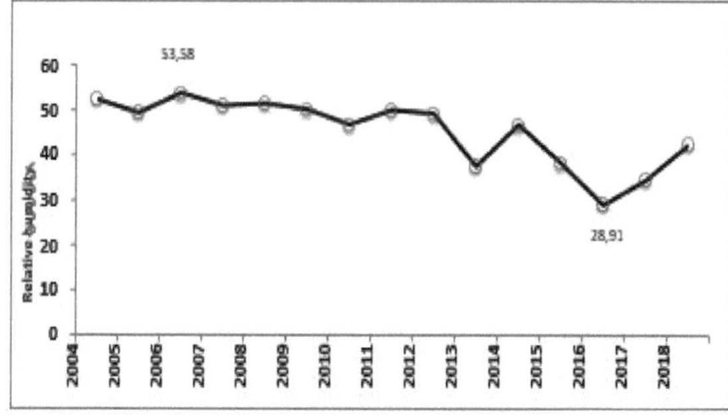

Figure 8: Interannual variation in relative humidity in the Laghouat region (2004-2018)

3.3.5 Jelly :

According to Le Houérou (1995), frosts are a limiting factor for agricultural practices and a constraining factor for natural vegetation. Frosts impose a cropping calendar that has to take account of the frost-free period, mainly for open-field vegetable crops and early-flowering arboriculture, which restricts their use to the hottest and least-watered seasons. As for natural vegetation, its growth is delayed, as it is closely linked to temperature (CENEAP, 2009). Our region experiences frequent and severe frosts in winter, with an average frequency of 17.73 days per year (Fig. 9).

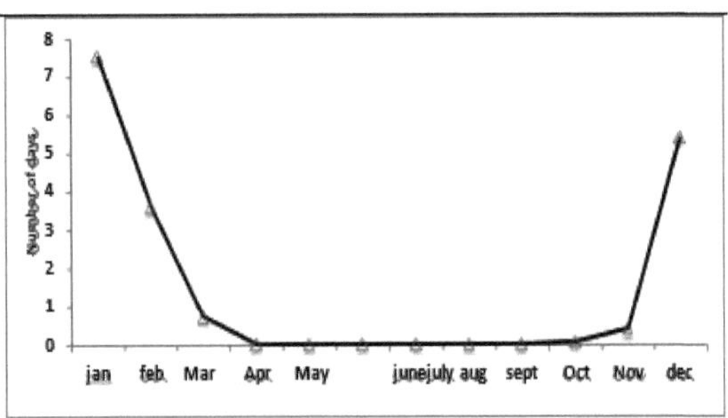

Figure 9: Average monthly number of days with frost in the Laghouat region (2004-2018)

The highest frost value was recorded in 2005, with 43 days, and the lowest was in 2016, with just 4 days (Fig. 10).

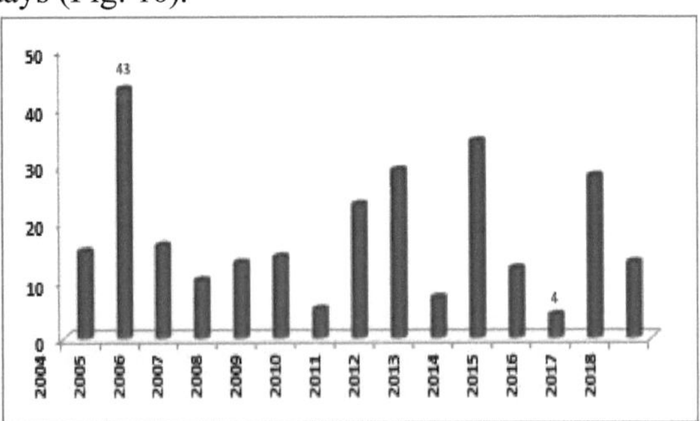

Figure 10: Interannual variation in the number of days with frost in the Laghouat region (2004-2018).

The highest values recorded in our orchard since its creation (2007) were 23 days in 2011, 29 days in 2012, 34 days in 2014 and 28 days of frost in 2017.

3.3.6 . Umbrothermal diagram :

The Bagnouls and Gaussen umbrothermal diagram is used in particular to highlight a possible period of biological drought in a given locality. It compares temperature and rainfall month by month. A period of the year is considered dry when the monthly rainfall, expressed in mm, is equal to or less than twice the temperature, expressed in degrees Celsius. The umbrothermal diagram is a classic way of presenting the climate of a region.
The umbrothermal diagram for the Laghouat region (2004-2018) shows a dry period covering the whole year (Fig. 11).

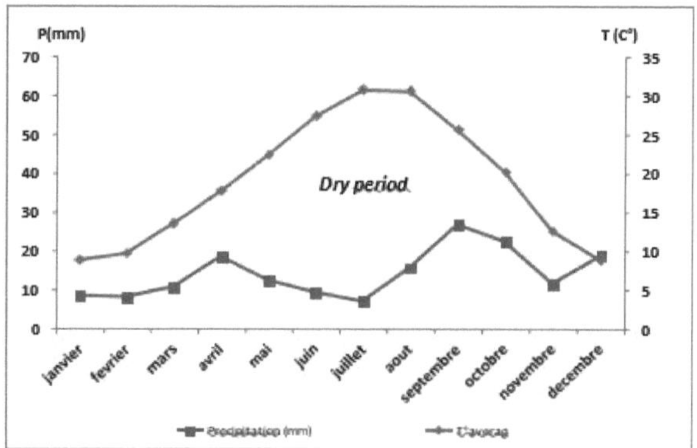

Figure 11: Bagnouls and Gaussen umbrothermal diagram for the Laghouat region (2004-2018).

3.3.7 . Rainfall-thermometer and Emberger climagram :

This quotient was set up by Emberger specifically to determine the types of Mediterranean climates, and is calculated using the following formula:

$Q2 = 2000*P/M^{2} - m .^{2}$

Q2: rainfall quotient.

P: average annual rainfall (mm). P(mm) = 167.83 mm.

M: average maximum temperature of the warmest month in degrees Kelvin. M= 37.22 °C.

m: average minimum temperature of the coldest month in degrees Kelvin. m = 2.3 °C.

T (°k) = T °C + 273.2.

Calculation of the Emberger quotient gave a value of 16.41, which places the region in a lower arid bioclimate, a thermal variant with cool winters (Fig. 12).

Laghouat

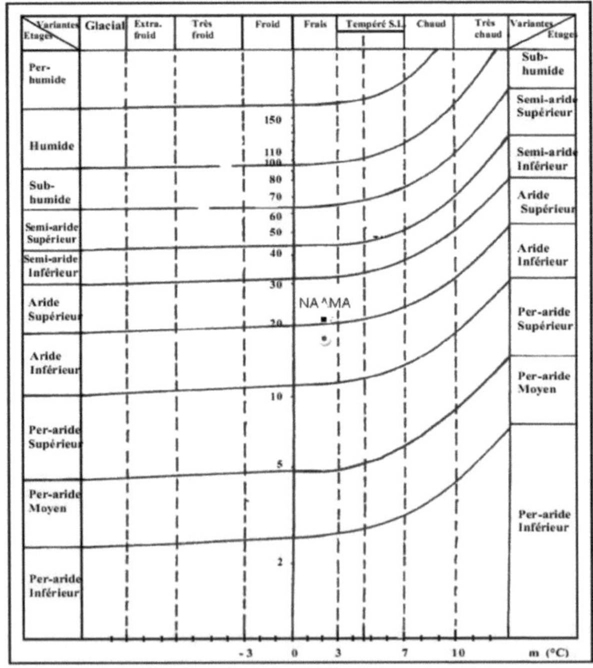

Figure 12: Emberger climagram for the Laghouat region (Daget, 1977)

3.3.8 . De Martonne aridity index :

This index is used to determine the degree of aridity of a region using the following formula:

$$\mathbf{I = P / T+10}$$

P: Total annual precipitation (**P=167**.83 mm).

T: Mean annual temperature (**T=18**.96°C).

Calculation of the De Martonne index gave a value of 5.79, which classifies the region as having an arid climate (Tab. 3).

Table 3: Aridity index (I) values and corresponding bioclimates :

Index value	Type of climate
0<I<5	Hyper-arid
5<I<10	Arid
10<I<20	Semi-arid
20<I<30	Sub-humid
30<I<55	Wet
I>55	Per-humid

3.4.Study site:

The site was chosen for its accessibility to work, located near the Bennacer Ben Chohra communal road, and fenced on its three other corners. The study olive grove is located 25 km south of Laghouat and covers an area of 25 ha (DSA, 2019). Our work was carried out between April and May 2019. We deliberately chose to carry out the sampling in an olive plot planted by the Laghouat forestry services (Tab. 4 and Fig. 13).

Table 4: Geographical coordinates of the study site

Website	Latitude	Longitude
Ben nacer ben chohra	33°39'0.49 "N	3° 10' E

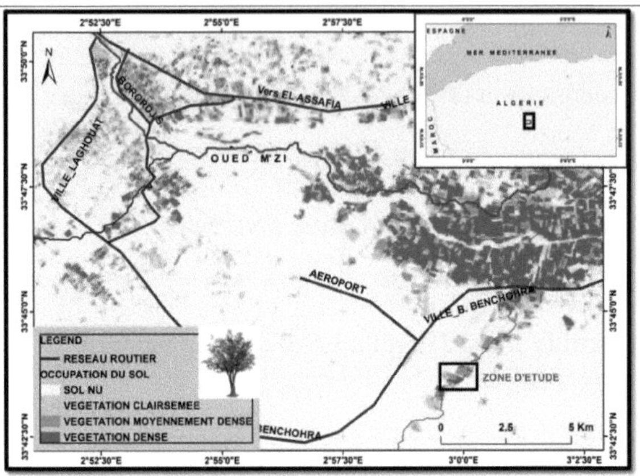

Figure 13: Geographical location of the study site.

The olive orchard is planted on a course of Anabasis articulata on skeletal soil with a bumpy limestone slab, where the limestone nodules are clearly visible (Photo 3).

Photo 3: Panoramic view of the study site (original)

The station is characterised by an average plantation density. The land is rocky and poorly maintained, the trees are young (12 years old) and poorly pruned (DSA, 2019) (Plate 2). The variety grown is Chemlal. This plot has already undergone phytosanitary treatment during our experiment (DSA, 2019). The planting distances applied are 5 m between trees and an average of 5 m between planting rows. The irrigation technique used is based on drip irrigation with an extremely degraded network where the water does not reach most of the trees.

3.5. Methodology

We found it useful to choose subjective sampling, which is defined by Gounot (1969) as "the simplest and most intuitive form of sampling". It consists of choosing samples that appear to be the most representative and sufficiently homogeneous (Long, 1974). Subjective sampling was used to select the study site, based on its accessibility and the availability of its planting history. Random sampling was used to select the lines and trees to be studied. We carried out observations and measurements on the olive tree between April and May 2019, at the height of the flowering period. All the measurements concerned the following parameters in order to characterise the olive tree population studied:

3.5.1. Success rate :

This is the ratio between the percentage of individuals planted and successful at the time of the assessment, and the number of individuals planted during the planting operation. Number of successful individuals

Success rate (%) =x 100

Number of individuals set up

The success rate is assessed in plots made up of the 103 closest individuals (Fig. 14).

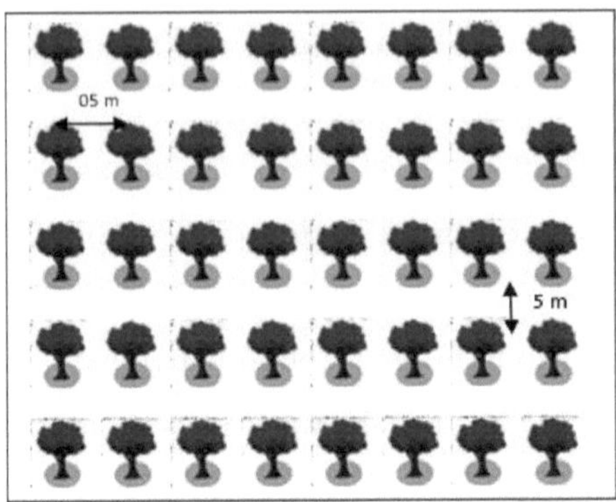

Figure 14: Schematic plan of the sampled site (original).

3.5.2. Tree dendrometric parameters

These parameters give us an idea of the state of development of the plantations. In our study, these parameters are limited to :

1. Height ;
2. Tree crown diameters (small and large).

3.5.2.1. Height of tree

Using a tape measure, we measured the height of the tree (Blozan, 2008). To do this, 102 trees were measured at the site (Photo 4).

Photo 4: Method for estimating height (H) and crown diameters D1 and D2

3.5.2.2. Shaft diameters

The large and small diameters were measured using a tape measure to characterise the population and get an idea of the volume that is the site of disease (Photo 4). Because of the shape of the tree, we had to measure two diameters to obtain an average diameter.

3.5.3. Leaf biometry

Measurements on the leaves concerned the length and width of the leaf as well as the length of the petiole. Using a caliper (0.01 mm), we took measurements on all the leaves studied (510 leaves), five at a time.
(5) leaves for each tree (Photo 5).

Photo 5: Technique for measuring olive leaf parameters

3.5.4. Damage and diseases observed

During the two field trips, various observations were made: the success rate of the plantation and the state of health of the individuals, which consisted of identifying pathogens or deficiencies according to the anomalies observed on site (leaf curling, leaf colour, pimpled leaves, shape of the leaf, etc.). limb). All observations concerned the leaves, branches and trunks. However, the period of our surveys did not coincide with the fruiting stage, which unfortunately prevented us from reporting any pests or diseases affecting this part of the plant. M. Amara and M. Kouidri were responsible for identifying the damage.

3.6. Statistical analysis

The analysis involved the descriptive statistics of all the parameters measured, with the mean, standard deviation and extremes reported (Avg ± Sd (min- max)) together with the coefficient of variation (CV%), which gives an idea of the homogeneity of the parameters measured. Significant correlations between the various parameters are also mentioned and interpreted. We used the Pearson correlation at a 95% confidence interval. These statistical analyses are obtained using **Statistix** 8 software under Windows and illustrations in **Excel** 2007.

PART 2

RESULTS AND DISCUSSION

CHAPTER 4
RESULTS

4.1.Success rate

Calculation of the success rate of olive tree planting in our orchard revealed a value of 99.02 %. In the orchard as a whole, we observed only two dead individuals, including one in our sample plot.

4.2.Biometric parameters

The main results obtained from measuring the biometric parameters of the 102 trees studied are shown in table (5).

Table 5: Main biometric parameters measured in the olive orchard

Parameter measured	Avg ±SD	Minimum	Maximum	Coefficient of variation
Total height in m	1,87 ± 0,29	1,10	2,75	15,81
Large diameter in m	2,24 ± 0,44	1,10	3,30	19,77
Small diameter in m	2,03 ± 0,42	1,00	3,20	20,69

4.2.1. Height of tree

The average height measured at the site was 1.87 ± 0.29 m, varying between 1.1 m and 2.75 m (Tab. 5). This represents a variation of 15.81%, which shows that the heights of all the individuals measured are similar.

4.2.2. Large diameter :

The large diameter recorded for all individuals is 2.24±0.44 m. It varies between 1.1 m and 3.30 m with a coefficient of variation (CV%) of 19.77% (Tab.5).

4.2.3. Small diameter :

The small diameter measured for all individuals is 2.03 ± 0.42 m. It varies between 1.00 m and 3.20 m with a coefficient of variation (CV%) of 20.69% (Tab.5).

4.3.Leaf biometry

All the measurements taken on olive leaves are shown in Table 6.

Table 6: Main leaf biometric parameters measured in the olive orchard

Parameter measured	Avg ±SD	Minimum	Maximum	Coefficient of variation
Sheet length cm	1,93 ± 0,23	1,32	2,74	12.15
Sheet width cm	0,25 ± 0,10	0,06	0,46	42,22
Length of petiole cm	0,08 ± 0,09	0,026	0,64	115%

4.3.1. Sheet length

The average leaf length measured at the site was 1.93 ± 0.23 cm, varying between 1.32 m and 2.74 cm in length (Tab. 6). This represents a variation of 12.15%.

4.3.2. Sheet width

The average width of the leaves measured was 0.25 ± 0.10 cm, ranging from 0.06 cm to 0.46 cm (Tab. 6). There is a large variation of 42.22%. We found a positive and statistically significant correlation between leaf width and length ($r=0.52$; ddl=501; $p\leq0.0001$) (Fig. 15). Leaf growth is in a normal physiological state.

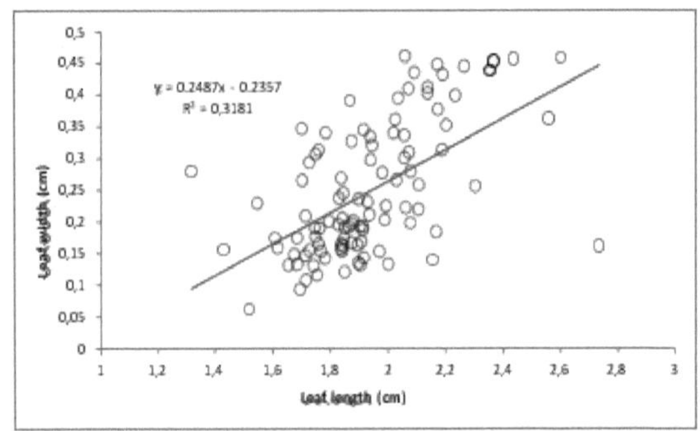

Figure 15: Relationship between leaf length and width

4.3.3. Length of petiole

The average length of petioles is 0.08 ± 0.09 cm, varying between 0.026 cm and 0.64 cm in length (Tab. 6). There is a very wide variation in measurement of the order of 100%, which reflects a high degree of heterogeneity in our measurements for this parameter. We found a positive and statistically significant correlation between petiole length and leaf width ($r=0.67$; ddl=501;$p\leq0.0001$) (Fig. 16).

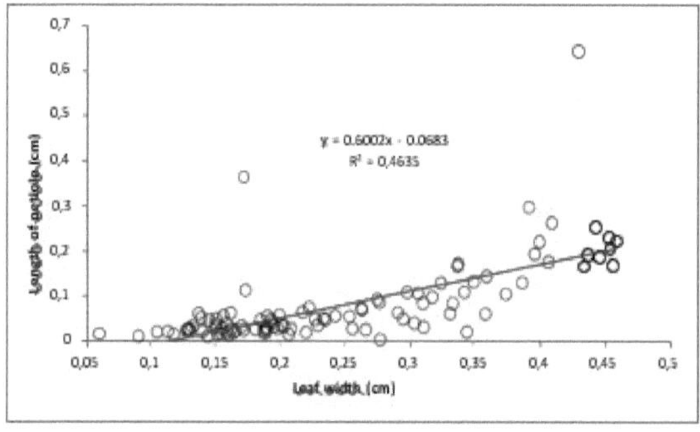

Figure 16: Correlation between leaf width and petiole length

There is also a positive and highly significant relationship between petiole length and leaf length (r=0.37; ddl=501; p=0.0015); longer leaves have longer petioles (Fig. 17).

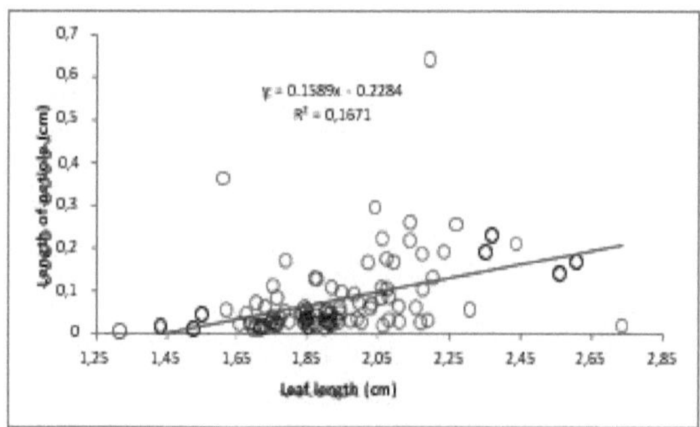

Figure 17: Correlation between leaf length and petiole length

4.4. Damage and disease parameters

The extent of damage and disease in the olive orchard studied is illustrated in the following figure.

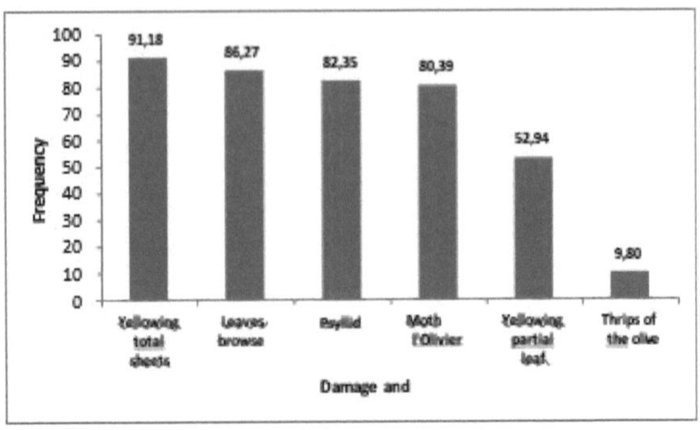

Figure 18: Main olive damage and diseases in order of importance in the study site

We found that there are two types of damage: physiological diseases caused by environmental factors (abiotic), such as drought or a lack of elements, and other diseases caused by pathogens or insects (biotic). In order of importance, these include :

a) Total leaf yellowing: This is the 1[er] phenomenon observed on olive trees in our orchard, with a rate of 91.18% of all trees (Fig. 18 and Photo 6). This phenomenon is the result of a lack of water for the plant;

Yellowing of leaves

Photo 6: Total yellowing of leaves (original)

b) Grazed leaves: caused by Otiorhyncus cibricollis, is the 2ème phenomenon affecting olive trees in our orchard, with an infestation rate of 86.27% of all trees. It consumes the leaves, making characteristic marginal indentations (Fig. 18 and Photo 7).

Photo 7: Damage caused by Otioryhucns cibricollis (browsed leaves)(original)

There is a positive and highly significant correlation between the presence of Otioryhucns cibricollis and the presence of Otioryhucns cibricollis. Otiorhyncus cibricollis and total leaf yellowing (r=0.25; ddl=101; p=0.03) (Fig. 19). Trees affected by Otiorhyncus cibricollis also suffer from water deficit.

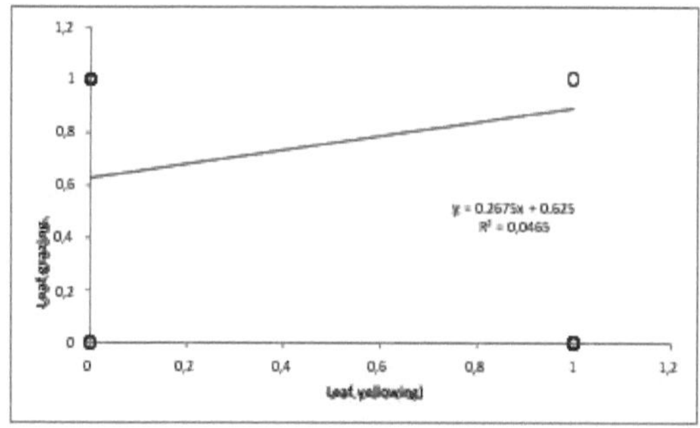

Figure 19: Relationship between leaf yellowing and Otiorhyncus cibricollis

c) Psyllid (Euphyllura olivina): This is the 3ème factor affecting olive trees in our orchard, with an infestation rate of 82.35% of all trees (Fig. 18 and Photo 8).

Photo 8: Olive psyllid (original)

d) Olive moth (Prays oleae): This is the 4ème factor influencing olive trees in our orchard with an infestation rate of 80.39% of all trees (Fig. 18 and Photo 9).

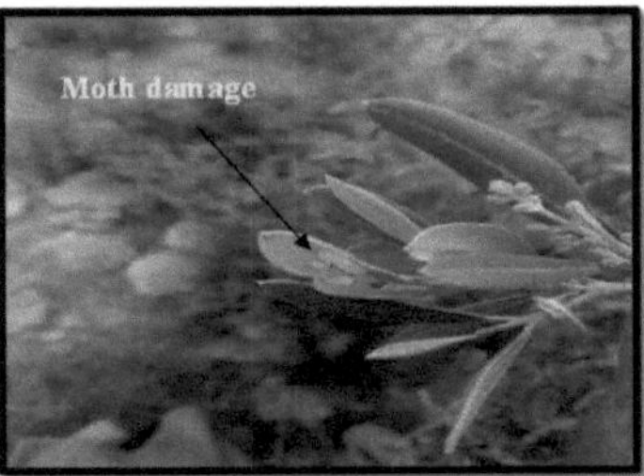

Photo 9:Moth damage (original)

There was a statistically significant positive correlation between leaf yellowing and olive moth (r=0.27; ddl=101; p=0.02); the trees most affected by the moth were also the most affected by leaf yellowing (Fig. 20).

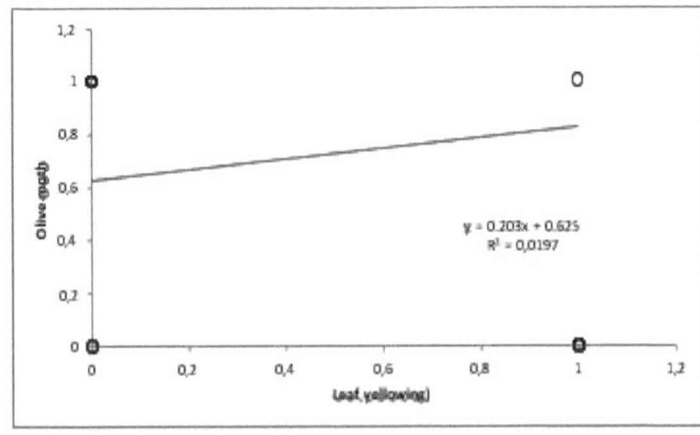

Figure 20: Relationship between olive moth and total leaf yellowing

The presence of the moth is also negatively correlated with leaf length (r=-0.33; ddl=101; p=0.0041) (Fig. 21).

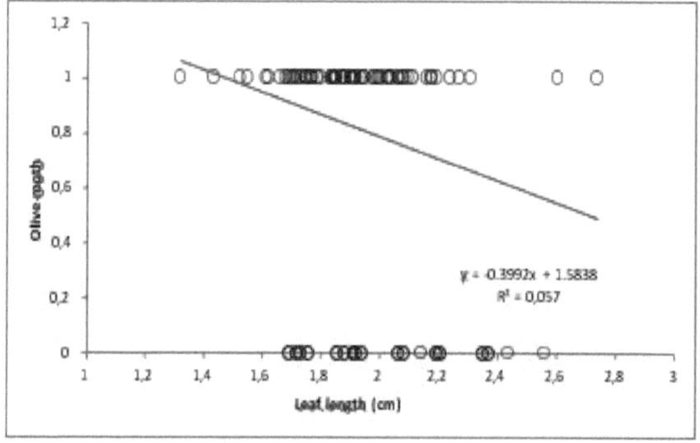

Figure 21: Relationship between olive moth and leaf length

Another negative relationship between olive moth and leaf width (r=- 0.37; ddl=101; p=0.0013) (Fig. 22).

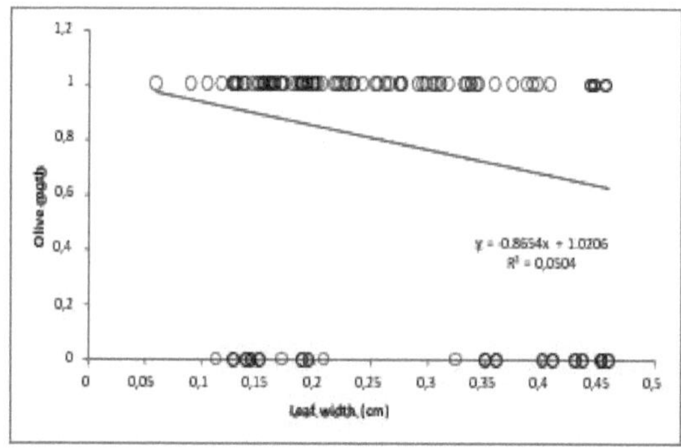

Figure 22: Relationship between olive moth and leaf width

There was also a negative relationship between Prays oleae and petiole length (r=-0.36; ddl=101; p=0.0017) (Fig. 23).

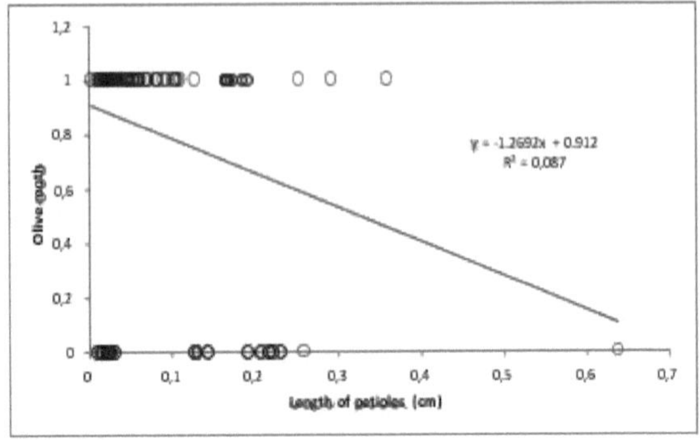

Figure 23: Relationship between olive moth and petiole length

The olive moth affects the youngest leaves, which are the weakest in length, width and petioles. The presence of the olive moth is positively correlated with the presence of Otiorhyncus cibricollis (r=0.48; ddl=101; p≤0.0001) (Fig. 24).trees that contain the moth also host Otiorhyncus cibricollis.

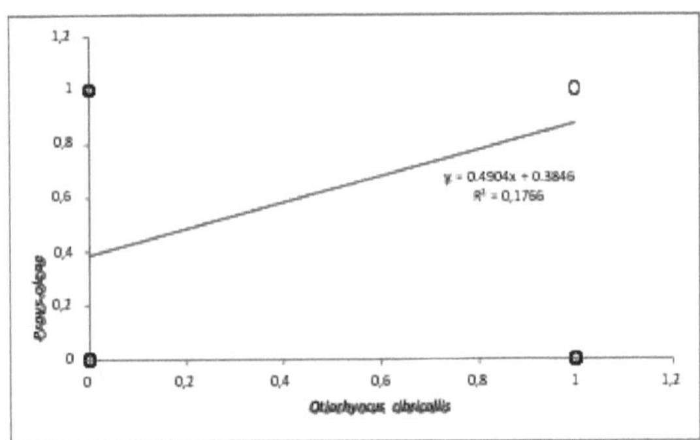

Figure 24: Correlation between the olive moth and Otiorhyncus cibricollis

e) Partial leaf yellowing: caused by potassium deficiency. This is the 5ème factor that characterizes olive trees in our orchard, with an infestation rate of 52.94% of all individuals (Fig. 18 and Photo 10).

Photo 10: Symptoms of potassium deficiency (partial yellowing) (original)

f) **Olive thrips** (Liothrips oleae)**:** This is the 6ème factor affecting olive trees in our orchard with an infestation rate of 9.80% of all trees (Fig. 18 and Photo 11).

Figure 11: Leaf deformation caused by olive thrips

CHAPTER 5

DISCUSSION

Olive trees, like all other tree crops, are attacked by a number of pests (insects, mites, fungi and bacteria) (Tahraoui, 2017). Tahraoui reports that 77 species of arthropods have been identified on olive trees in western Algeria. These include pests such as Saissetia olea, the pollinator Apis mellifera and predators such as Pnigaleo mediterraneus. Hosni (2006) reports the presence of 14 insect species on the olive tree. 5 of these are pests such as Psylla oleae, Saissetia oleae and Bactrocera oleae. The other pests are predators such as Coccinella septempuctata and Chrysopa vulgaris. Part (1997) points out that more than 60 known species of insect live on the olive tree in the Mediterranean, of which around 15 to 20 species are permanent or occasional parasites. Cautero (1965) in Gaouar (1996) lists nearly 250 major enemies reported by various authors. These include 90 fungi, 5 bacteria, 3 lichens, 4 mosses, 3 angiosperms, 11 nematodes, 110 insects, 13 arachnids, 5 birds and 4 mammals. Daane et al (2005) report that, in Algeria, the black scale Saissetia oleae, the mealybug Pollinia pollini and the violet scale Parlatoria oleae are the most common and most serious pests of the olive tree. Coutin (2003) points out that the pests that cause the most damage to olive trees are the olive fly (Bactrocera oleae), the black scale insect (Saissetia oleae) and the moth (Prays oleae). In the present study, we encountered the following pests, in order of importance, Olive Otiorhynchus (Otiorhyncus cibricollis), Psyllid (Euphyllura olivina), Olive Moth (Prays olae), and Olive Thrips (Liothrips oleae). However, Hobaya and Bendimerad (2012) in Tlemcen, cited the main pests in order of importance of damage, Saissetia oleae (40%), Bactocera oleae (25%), Sturnus vulgaris (15%), Liothrips oleae (15%), Otiorhyncus sp. (5%) and Cribricolis sp. (5%). The findings of this last author are the result of several seasons of monitoring. The absence of some pests can also be explained by the ecological position of the study station; many authors point out the importance of stations and their impact on the infestation rate. Indeed, Delrio and Cavalloro (1977) and Jerraya et al (1982) state that attacks are greater in coastal areas than in the interior of countries. Gaouar (1989) reports that severe drought aggravated by very frequent siroccos reduces the rate of infestation by these pests.

Olive fly

According to Bonnemaison (1962) and Hobaya and Bendimerad (2012), the olive fly is the main pest of the olive tree, and one of the most feared parasitic insects in olive groves. Hmimina (2009) points out that in its development cycle, the first flies fly early (February to March), but if there are no olives, they die without reproducing. This would explain the absence of the latter during the period of our sampling.

Black scale insect

Alford (1994) points out that the black scale insect (Saissetia oleae) is one of the main pests of the olive tree. It does not cause direct damage like the fly or the moth, but it can cause very serious weakening of the trees affected. However, we have not encountered this species in our orchard.

Olive moth :

According to Jardak et al (2000), the moth is the first major pest to be observed under the leaves of olive trees in March. This pest can cause significant crop losses. According to Loussert and Brousse (1978) and Jardak et al (2000), the development cycle is a succession of three generations. Damage takes the form of sinuous galleries. The second generation Known as anthophagous, the moths that hatch from the caterpillars of the first generation lay their eggs on the calyx of flower buds in April-May, after which the caterpillars hatch (Loussert and Brousse, 1978). According to Bonifacio (2009), it is the caterpillars that cause all the damage. The caterpillars of the 1ère generation feed on the flower buds, causing problems with fertilisation and fruit set. In our case, the moth (Prays olae) is the third most common pest of olive trees, after the olive moth (Otiorhyncus cibricollis) and the olive psyllid (Euphyllura olivina).

Olive psyllid

According to Jardak et al (1984), the development of psyllids results in characteristic spectacular symptoms (cottony clusters, honeydew and wax). The resulting damage in the event of high population density is firstly direct, causing abortion of the flower bunches or their wilting and dropping, leading to a reduction in the fruit set rate, and secondly indirect, causing a weakening of the plant through the development of fumagine following the secretion of honeydew by the larvae. However, Ksantini (2003) points out that it is important to apply appropriate pruning to aerate the tree and, in particular, the flower clusters. Unfortunately, this action is totally absent in our orchard, which has not undergone any pruning.

Olive Otiorhynchus

According to INRA (2010), the biology of Otiorhyncus shows that it is infaunal to the Olive tree, but highly polyphagous. The adult commonly attacks fruiting Rosaceae, Citrus, Cotton and Artichoke. The larvae live off the roots of alfalfa and mugwort (Artemisia). The adults, which appear at the end of May, are nocturnal. I.N.P.V. (2010), stresses that the damage caused by larvae is insignificant compared to that caused by adults. On olive trees, the leaves are cut by notches around their edges. During exceptional outbreaks, the attack can result in total defoliation. Pala et al (1997) report that climatic conditions (high relative humidity, mild temperatures) combined with a lack of maintenance under the trees, particularly in intensive, irrigated plantations, favour the multiplication of Otiorrhynche.

Thrips

According to Hmimina (2009), Thrips are insects measuring 1 to 2 mm in length that bite plant organs to feed on the cell contents. Duriez (2001) reports that the foliage of affected plants is marked with tiny grey spots, taking on the appearance of silver streaks over time. Young shoots, flowers and fruit become deformed, then necrotic, and the leaves eventually dry out. Thrips are tiny, discreet insects that are difficult to observe. Thrips can weaken the tree and transmit viral diseases, such as tomato brown spot (which can affect many plants). According to Civantos (1995), prevention is based on a simple principle: humidify. Thrips do not develop when there is sufficient humidity. He points out that integrated biological control is also of interest. Certain bugs (several species of the Orius genus), certain mites (such as Amblyseius cucumeris) and a nematode (Steinernema feltiae) are natural predators of Thrips. It should be noted that the organs most affected are the leaves and inflorescences in our case. Generally, leaves and fruit are the most targeted by pests (Afiol, 2001; Al Ahmed and Al Hamidi, 1984; Hobaya and Bendimerad, 2012).

Deficiencies

Leaves are also good indicators of the tree's state of health; the total or partial yellowing of the leaves observed in the olive trees in our study would seem to be the repercussion of the influence of a physiological problem in these trees. Partial or total yellowing, followed by leaf drop, is thought to be the result of poor irrigation, which has already been observed during our outings. According to Belguerri (2016) and Dib (2017), in the case of potassium deficiency in olive trees, symptoms appear on the leaves and begin with chlorosis of the apical part. Leaf discolouration progresses towards the base, giving the leaf blade a bronze colour. Potassium regulates the plant's water metabolism under drought conditions (water stress) (Dib, 2017). Apical leaf chlorosis can be confused with boron deficiency (it has been observed in the field); the latter only affects the leaf tips. Boron availability in the soil decreases under drought conditions and on soils with a high pH, particularly calcareous soils (Belguerri, 2016). These latter problems require chemical analyses of the soil and leaves to help resolve this type of deficiency.

CONCLUSION

Our study was carried out between April and May 2019 in an olive orchard located south of Laghouat. The results revealed that the trees suffer from several problems of abiotic and other biotic origin. Measurements taken on individual olive trees (102 trees) show a success rate of 99.02% and that their average height after 12 years of planting does not exceed 1.87 ± 0.29 m. Their diameters averaged 2.24 ± 0.44 m and 2.03 ± 0.44 m respectively.± 0.42m for the large and small diameters.Leaf biometry shows an average length of 1.93 ± 0.23 cm and an average width of 0.25 ± 0.10 cm with a petiole 0.08 ± 0.09 m long. The main pests observed during the study period were the olive otiorhynchid (86.27% infestation rate), the olive psyllid (82.35%) and the olive moth (80.39%), which attacks the youngest leaves. We noted the absence of olive fly and black scale, which are generally the most dreaded pests in Algerian olive plantations.The symptoms observed on the leaves of the olive trees studied indicate a persistent drought (91.18% of trees with yellow leaves) and a possible deficiency in some elements (potassium, boron, etc.), which means that a fertilisation assessment needs to be carried out to correct the situation in this orchard. This study is an initial contribution to assessing the state of health of olive tree-based development plantations. It needs to be monitored throughout the year, and on other plots and other species. It is a valuable tool in the hands of decision-makers, making it easier to take sound management decisions on projects of economic interest.

REFERENCES

Alford D. V., 1994. Pests of ornamental plants - French version. Ed. INRA, 464 p.

Argerson C., 2008. Olive growing in the world, production and trends, le nouvel Olivier, p. 11.

Artaud M., 2008. L'olivier sa contribution dans la prévention et le traitement du syndrome métabolique. Ed. C.O.I., 30 p.

Assawah M.W., Ayat M., 1985. On certain diseases of olive trees at Oran area. Premières Journées Scientifiques de la Société Algérienne de Microbiologie. April, Institut Pasteur, Algiers, Algeria,

Bardbury, J. F., 1986. Guide to Plant Pathogenic Bacteria. CAB International Mycologican Institute. Kew, UK. 332p.

Belguerri H., 2016. Contribution to the study of the effect of irrigation and nitrogen and potassium fertilization on the productive and qualitative performance of the super-intensive olive tree. Doct. thesis Univ. of Lleida, 166p.

Belhoucine S., 2003 - Etude de l'éventualité d'un contrôle biologique contre la mouche de l'olivier dabs cinq stations de la wilaya de Tlemcen. Magister's thesis, Univ. Tlemcen, 94 p.

Bellahcene M, 2004. La verticilliose de l'olivier: étude épidémiologique et diversité génétique de Verticillium dahliae Kleb, agent de la verticilliose. PhD thesis, University of Oran, Algeria, 145 p.

Bellahcene M.; Fortas Z.; Fernandez D.; Nicole M., 2005a. Vegetative compatibility of Verticillium dahlia isolated from olive trees (Olea europea L.) in Algeria. Afric. J. Biotechn. 4(9): 963-967.

Bellahcene M., Assigbetsé K., Fortas Z., Geiger J.P., Nicole M., Fernandez

D., 2005b. Genetic diversity of Verticillium dahliae isolates from olive trees in Algeria. Phytopathol. Mediterr. 44: 566-274.

Berlioz C., 1950 - Les produits agrochimiques en oléiculture et leur impact sur l'environnement. Olivæ, n°65, p.p. 32-39.

Boutkhil S., 2012. The main fungal diseases of olive (Olea europea) in Algeria: geographical distribution and importance. Thesis Mag. Univ. Oran, 133 p

Brikci N., 1993. Effectiveness of an optimised insecticide treatment on the olive pest Dacus oleae in the Tlemcen region. Mémoire D.E.S biologie,

CIHEAM, 1988. L'olive de table comme production alternative à la production d'huile d'olive pp. 187 - 191.

Coutin R., 2003 - Les insectes de l'olivier. Insectes, 1 9 (3) : 1 3 0.

Daget P. H, 1988. Un élément actuel de la caractérisation du monde méditerranéen: de la liaison du PNTTA. N°152.p1. Ecole nationale d'agriculture, Meknès,Kingdom of Morocco.Amouretti and Comet, 2000

Diab N. and Deghiche L., 2014. Arthropods present in an olive crop in Saharan regions case of El outaya plain , tenth international conference on pests in agriculture department of agronomy, Mohamed Kheider university. Biskra. Algeria 6p

Direction des Services Agricoles de la wilaya de Laghouat (D.S.A.), 2019. Report on olive tree planting in the Laghouat region. 2p.

Djadoun S., 2011, Influence de l'hexane acidifient sur l'extraction d'huile de grignon d'olive assistée par microondes. Master's thesis, Univ. Mouloud Mammeri, Tizi Ouzou, 87 p.

Duriez J. M., 2001. Guide du planteur d'oliviers. Ed. Laguedoc-Roussillon, 22 p.

Gadan, L., Bollet, C., Abu Ghorrah, M., Grimont, F. and Gimont, P. A. 1992. DNA elatedness among the pathovarstrains of Pseudomonas syingae subps. savastanoi. Janse (1982) and poposal of Pseudomonas savastanoi sp. Nov. Internatial Jiurnal of Systematic Bacterioilogy. **42** :606-12.

Gaouar N. and Debouzie D., 1991 - Olive fruit fly Dacus oleae Gmel. (Diptera- Tephritidae) damage in Tlemcen region, Algeria. J. App. Ent, (112): 288-297

Gaouar-Benyelles N., 1996 - Apport de la biologie des populations de la mouche de l'olivier Bactrocera (Dacus) oleae Gmel (Ditera: Tephiritdae) à l'optimisation de son contrôle dans la région de Tlemcen. Doctoral thesis. University of Tlemcen, Algeria, 116 p.

Gazeau G., 2012. Fertilisation des Olivers (9): 4p.

Henry S., 2003. L'huile d'olive, son intérêt nutritionnel, son utilisation en pharmacie et en cosmétique. Diplôme d'Etat de docteur en pharmacie, Univ. Henri Poincaré, 127 p.

Hmimina M., 2009a. Olive fly, technology transfer in agriculture.(183) : 4.p

Hmimina M. 2009b. The main pests of the olive tree, the fly, the moth, the psyllid and the black scale. Bull. Men. Inf. et Liaison du PNTTA, 4 p.

Hobaya O. and Bendimerad M., 2012. Contribution à l'étude des ravageurs de l'Olivier Olea europeae à Tlemcen. Mémoire d'ingénieur d'Etat en Agronomie, University of Tlemcen, 78 p.

Hosni A., 2006. Inventory of the pest in some perennial olive and citrus crops, specific study of the infestation rate in the Tlemcen region. Agronomy engineering dissertation, Univ. Tlemcen, 76 P.

Iguergaziz N., 2012. Trial to develop a food in the form of whole and/or de-sweetened date tablets with added aqueous extract of Algerian olive leaves.

Thesis of Mag. Univ. M'hamed Bougara, Boumerdas, 129 P.

I. N. P. V. (Institut National de la Protection des Végétaux), **1994**. Fiche technique des ennemis de l'olivier pour les différents stades, 5 p.

I. N. P. V. (Institut National de la Protection des Végétaux), **2009**. Technical file on Bactocera oleae, 2 p..

Jardak T., Jarraya A., Ktari M. and Ksantini M., 2000. Modelling trials on the olive moth, Prays oleae (Lepidoptera, Hyponomeutidae). Olivæ, (83): 22 Ŕ 26.

Kasraoui F., 2010. Practical study of olive tree requirements. Mémoire d'ing. Agronomie, Univ. Tlemcen 40 P.

Khris B., 2013. Agriculture : performances pour l'oléiculture et l'agrumiculture en 2013, Rédaction radio net, 2 p.

Labaali K., 2009. Chemical characteristics of olive tree soil at the end of flowering and beginning of fruit set. Mémoire ing. Agronomie, Univ. Cadi Ayad, Marrakech, 56 p.

Loussert and Brousse G., 1978. L'olivier technique agricole et production Méditerranéenne. Ed. Maisonneuvre et Lorose, 468 p.

Masterson P., 2007. Olive pests and their control in the Near East. FAO document, 178

Mary, B., Recous, S., Darwis, D., and Roben, D. (1996). Interactions between decomposition of plant residues and nitrogen cycling in soil. Plant and soil 181: 71.82.

National Office of Meteorology (ONM). (2019). Meteorological data of Laghouat. 7p

Pascal F. et Peris N., 1992- Les produits agrochimiques en oléiculture et leur impact sur l'environnement. Olivæ, (65) : 32-39p., 606 p.

Philippe, L. 2007. Plant pathogenic prokaryotes. Chapter 4 phytopathologies. Ed Feeman, New York. 432p.

Rahmani M., 1989. The role of chlorophyll pigments in the photooxidation of virgin olive oil. Olivae, (26): 30 - 32.

Rahmani A et al., 1998. Contribution à l'élaboration d'un schéma directeur d'assainissement en milieu rural. Commune de Terny (Wilaya DE Tlemcen). Ing. Hydr.Univ.Tlemcen.82p

Saad D., 2009. Study of the endomycorrhizae of the Sigoise variety ofolive tree (Olea europea L.) and trial of their application to semi-woody cuttings. Mém. Magister, Univ. Oran, 98p.

Sasanelli, N., 2009. Olive nematodes and their control, CNT Plant Protection Institute. Management of fruit corps and forest nematodes. 275-315.

Sekour B., 2012. Phytoprotection de l'huile d'olive vierge (H.O.V) par ajout des plantes végétales (thym, ail, romarin). Magister thesis, Univ .M'hamed Bougara, Boumerdas,110 p.

Tahraoui A., 2015. Inventory of the entomological fauna associated with olive trees in the Tlemcen region. Mém. Master , unvi. Tlemcen p55.

Printed by Books on Demand GmbH, Norderstedt / Germany